Physics Poem 101

DIPAN KUMAR DAS
SUDIP KUMAR DAS

"Reflection"

Reflection, in still waters deep,
A mirror to our souls, a secret to keep.
A glimpse of truth, beyond what we see,
A window to the past, and all we used to be.

The ripples of our actions, in a dance so true,
A symphony of choices, the story of me and you.
Each wave a lesson learned, a memory we've made,
A marker of our journey, through life's winding braid.

In the depths of reflection, we find a clearer view,
A chance to understand, what we're meant to do.
To heal the wounds that linger, to forgive and to grow,
To face the fears that hold us, and let our spirit show.

So let us embrace reflection, and all that it can bring,
A journey of self-discovery, to help us spread our wings.
And rise above the surface, to face a brighter day,
With confidence and courage, to lead us on our way.

"Refraction"

Refraction, a bending of light,
A dance of colours, so vibrant and bright.
A spectrum of hues, a rainbow so fair,
A glimpse of magic, beyond what's rare.

It's the play of light, through raindrops and glass,
A bending of beams, a moment that lasts.
A prism of beauty, a work of art,
A window to wonder, that touches the heart.

It's the mystery of physics, a lesson so grand,
A study of nature, with knowledge so vast and so planned.
A demonstration of science, a marvel so pure,
A reminder of how much, the world still has in store.

So let us marvel at refraction, a dance so bright,
A symphony of colours, that fill us with delight.
And bask in the beauty, of this wondrous sight,
As we marvel at nature, and all its might.

Snell's law

Snell's law, a principle so true,
A guideline for light, that shines its way through.
An equation so simple, yet oh so grand,
A theory that explains, the motion of light through our
land.

It's the bending of beams, as they cross a divide,
A shift in direction, that cannot be denied.
A result of refraction, and the speed of the light,
A force of nature, that guides it just right.

It's the foundation of optics, a key to our sight,
A principle that shapes, the world we observe in its light.
From lenses to prisms, it guides our view,
A cornerstone of science, that brings us wisdom too.

So let us embrace Snell's law, and all that it brings,
A path to discovery, and a world of wondrous things.
And in the bending of light, let us find our way,
To a brighter tomorrow, and a world of endless day.

interference

Interference, a dance of waves,
A symphony of light, that our eyes cannot save.
A union of crests, a harmony so grand,
A display of beauty, in the touch of a hand.

It's the collision of beams, that form a bright line,
A moment of wonder, that captures the mind.
A result of the interference, of two sources so bright,
A demonstration of physics, that shines so bright.

It's the pattern of light, that dances so free,
A tapestry of colours, that's woven in thee.
From Young's slits to double-slits, it's a sight to behold,
A principle of nature, that brings magic to unfold.

So let us embrace interference, and all that it brings,
A journey of discovery, and a world of wondrous things.
And in the harmony of waves, let us find our way,
To a brighter tomorrow, and a world of endless day.

Young's Double-Slit Experiment

Young's Double-Slit Experiment, a mystery so grand,
A demonstration of light, that struck a chord in our land.
A test of waves or particles, a question to unfold,
A principle of physics, that left scientists bold.

It's the beams of light, that pass through two slits so
fine,
A dance of interference, that forms a pattern divine.
A demonstration of duality, a mystery so real,
A principle that shaped, the way we see and feel.

It's a test of our limits, a challenge so bright,
A glimpse of the universe, that's waiting for us to shed
light.
From photons to electrons, it's a story untold,
A principle of nature, that never grows old.

So let us embrace Young's experiment, and all that it
brings,
A journey of discovery, and a world of wondrous things.
And in the duality of light, let us find our way,
To a brighter tomorrow, and a world of endless day.

Polarization

Polarization, a twist of light,
A dance of waves, that's out of sight.
A filter of colours, a shimmer so bright,
A principle of physics, that illuminates the night.

It's the orientation of light, that shines so bright,
A beam of energy, that's seen in the light.
A demonstration of direction, a mystery so grand,
A principle of nature, that gives us a helping hand.

It's the polaroid's that filter, the light that we see,
A guide to the rainbow, a spectrum so free.
From the sun to the sky, it's a story untold,
A principle of physics, that's waiting to unfold.

So let us embrace polarization, and all that it brings,
A journey of discovery, and a world of wondrous things.
And in the twist of light, let us find our way,
To a brighter tomorrow, and a world of endless day.

Diffraction

Diffraction, a bend of light,
A dance of waves, that's out of sight.
A spread of colours, a shimmer so bright,
A principle of physics, that illuminates the night.

It's the spreading of light, as it passes a gap,
A demonstration of waves, that makes us clap.
A phenomenon so real, a mystery so grand,
A principle of nature, that guides us by the hand.

It's the grating of light, that splits the beam,
A display of diffraction, a sight so serene.
From light to sound, it's a story untold,
A principle of physics, that's waiting to unfold.

So let us embrace diffraction, and all that it brings,
A journey of discovery, and a world of wondrous things.
And in the bend of light, let us find our way,
To a brighter tomorrow, and a world of endless day.

Refractive index

Refractive index, a measure so grand,
A guideline for light, that travels the land.
A number so simple, yet full of might,
A principle of optics, that guides our sight.

It's the speed of light, as it travels through air,
A measure of change, that's beyond compare.
A ratio of velocities, a story untold,
A principle of nature, that never grows old.

It's the bending of light, as it crosses a line,
A demonstration of refraction, that shines so bright and fine.
From lenses to prisms, it guides our view,
A cornerstone of science, that brings us wisdom too.

So let us embrace the refractive index, and all that it brings,
A journey of discovery, and a world of wondrous things.
And in the speed of light, let us find our way,
To a brighter tomorrow, and a world of endless day.

Total internal reflection

Total internal reflection, a phenomenon so rare,
A bend of light, that dances with care.
A reflection so complete, it's hard to comprehend,
A principle of optics, that's beyond our grasp and bend.

It's the light that bounces, within a boundary so bright,
A demonstration of physics, that shines so bright and right.
A reflection so total, it's a mystery to see,
A principle of nature, that's waiting to be free.

It's the fibers of light, that bring us the internet,
A network of communications, that we can't forget.
From optics to medicine, it's a story untold,
A principle of science, that never grows old.

So let us embrace total internal reflection, and all that it brings,
A journey of discovery, and a world of wondrous things.
And in the bend of light, let us find our way,
To a brighter tomorrow, and a world of endless day.

Microscope

The microscope, a tool so grand,
A window to the world, that helps us understand.
A lens of discovery, that shines so bright,
A device of science, that brings us new sight.

It's the magnification of details, that we cannot see,
A demonstration of optics, that sets us free.
A tool for science, a mystery so real,
A device that's waiting, to reveal what we feel.

It's the cells that multiply, the germs that we fight,
A glimpse into the world, that's hidden from sight.
From biology to chemistry, it's a story untold,
A device that brings us, knowledge and bold.

So let us embrace the microscope, and all that it brings,
A journey of discovery, and a world of wondrous things.
And in the lens of science, let us find our way,
To a brighter tomorrow, and a world of endless day.

Astronomical telescope

The astronomical telescope, a tool of the night,
A window to the stars, that shines with all its might.
A lens of discovery, that reaches to the sky,
A device of science, that makes us wonder why.

It's the distant galaxies, the planets that we see,
A demonstration of optics, that sets our minds free.
A tool for astronomy, a mystery so vast,
A device that's waiting, to reveal what will last.

It's the birth of stars, the death of black holes,
A glimpse into the universe, that never loses control.
From astrophysics to cosmology, it's a story untold,
A device that brings us, knowledge and bold.

So let us embrace the astronomical telescope, and all
that it brings,
A journey of discovery, and a world of wondrous things.
And in the lens of science, let us find our way,
To a brighter tomorrow, and a world of endless day.

Aberrations

Aberrations, a flaw in sight,
A hindrance to optics, that shines so bright.
A deviation from perfection, that can't be denied,
A challenge to science, that waits to be tried.

It's the distortion of images, that's hard to rectify,
A demonstration of physics, that tests our might.
A flaw in the lens, that affects what we see,
A challenge to optics, that's waiting to be free.

It's the chromatic aberration, that divides the light,
A hindrance to colours, that cannot be right.
From telescopes to microscopes, it's a story untold,
A challenge to science, that never grows old.

So let us embrace the aberration, and all that it brings,
A journey of discovery, and a world of wondrous things.
And in the quest for perfection, let us find our way,
To a brighter tomorrow, and a world of endless day.

Dispersion

Dispersion, a rainbow in sight,
A spectrum of colours, that shines so bright.
A separation of light, that can't be denied,
A principle of optics, that's waiting inside.

It's the bending of light, that creates the hue,
A demonstration of physics, that shines in view.
A separation of colours, that's hard to comprehend,
A principle of nature, that's waiting to bend.

It's the spectrum of light, that we see in the sky,
A demonstration of science, that cannot deny.
From raindrops to prisms, it's a story untold,
A principle of optics, that never grows old.

So let us embrace the dispersion, and all that it brings,
A journey of discovery, and a world of wondrous things.
And in the rainbow of light, let us find our way,
To a brighter tomorrow, and a world of endless day.

Light Emitting Diode

The Light Emitting Diode, a marvel of light,
A source of illumination, that shines so bright.
A tiny device, that brings us energy,
A tool of science, that sets us free.

It's the power of electrons, that creates the glow,
A demonstration of physics, that we come to know.
A device that's waiting, to be our guide,
A tool of technology, that's always at our side.

It's the light that shines, in our homes and streets,
A demonstration of engineering, that can't be beat.
From lighting to displays, it's a story untold,
A device that brings us, knowledge and bold.

So let us embrace the LED, and all that it brings,
A journey of discovery, and a world of wondrous things.
And in the light of science, let us find our way,
To a brighter tomorrow, and a world of endless day.

Laser.

A beam of light, so precise and bright,
A tool of science, with power untold.
The laser, a marvel of modern might,
A shining star, in the world of technology old.

From medicine to communication,
It's a tool that finds application,
With the power to cut, to heal,
It's a force that always appeals.

Its beams so narrow, so concentrated,
A ray of light, that's never faded,
A master of precision, a ruler of speed,
A tool that brings power, with every deed.

So here's to the laser, a shining light,
A device that brings hope, and banishes night,
A symbol of progress, a sign of our time,
A tool that will always, continue to shine.

Laser hazards.

A beam of light, so powerful and bright,
A tool of science, with potential to ignite.
The laser, a marvel of modern might,
A double-edged sword, in the world of technology
bright.

It brings us power, it brings us sight,
But in the wrong hands, it can cause fright.
With a beam that's so intense, so strong,
It can do harm, where it does not belong.

Its power so concentrated, so precise,
Can cause injury, if not treated with vice,
A hazard to the eyes, a risk to the skin,
A danger to humanity, if not handled within.

So here's to caution, with this beam of light,
A reminder that we must treat it with fright,
A symbol of progress, a sign of our time,
A tool that we must handle, with utmost care and
prime.

Radiation

Radiation, a force both bright and warm,
A source of energy, that can cause alarm.
A mysterious power, that travels far,
A journey of discovery, that takes us to the stars.

It's the heat that warms us, the light that guides,
A principle of physics, that cannot be denied.
A tool that's waiting, to be understood,
A journey of science, that's always good.

It's the energy that drives, our world so vast,
A demonstration of nature, that's unsurpassed.
From stars to sun, from galaxies to cells,
A principle of existence, that always compels.

So let us embrace the radiation, and all that it brings,
A journey of discovery, and a world of wondrous things.
And in the force of energy, let us find our way,
To a brighter tomorrow, and a world of endless day.

CT Scans.

CT Scans, a journey through the body,
A tool of medicine, that brings us to glory.
A glimpse inside, with every scan,
A story of health, and what's within.

It's the power of X-rays, combined with technology,
A tool that's waiting, to set us free.
A source of knowledge, that helps us heal,
A journey of discovery, that's always real.

It's the window to the inside, that we seek,
A demonstration of science, that's always unique.
From bones to organs, it gives us sight,
A tool that brings us, knowledge and light.

So let us embrace the CT Scans, and all that they bring,
A journey of discovery, and a world of wondrous things.
And in the power of medicine, let us find our way,
To a healthier tomorrow, and a brighter day.

Coulomb's law.

Coulomb's law, a principle of might,
A force that governs, electrons and light.
A rule of attraction, and repulsion too,
A journey of physics, that's always true.

It's the power that holds, the universe in place,
A demonstration of science, with an elegant grace.
From atoms to stars, it's always at work,
A tool that brings us, knowledge and perk.

It's the force that balances, the charge in space,
A journey of discovery, with a steady pace.
From lightning to magnets, it guides our way,
A principle of existence, that never goes astray.

So let us embrace Coulomb's law, and all that it brings,
A journey of physics, and a world of wondrous things.
And in the force of attraction, let us find our way,
To a deeper understanding, of the universe each day.

Ohm's law.

Ohm's law, a principle of wire,
A force that governs, the flow of fire.
A rule of resistance, and voltage too,
A journey of electricity, that's always true.

It's the power that drives, the circuits we build,
A demonstration of engineering, with a steady pulse and thrill.
From lamps to motors, it's always at work,
A tool that brings us, energy and perk.

It's the force that balances, the flow of energy,
A journey of discovery, with a simple harmony.
From transformers to generators, it guides our way,
A principle of electronics, that never goes astray.

So let us embrace Ohm's law, and all that it brings,
A journey of electricity, and a world of wondrous things.
And in the force of resistance, let us find our way,
To a brighter future, with technology each day.

Transistors.

Transistors, a wonder of the age,
A tool of technology, that turns the page.
A switch of signals, and amplifier too,
A journey of electronics, that's always new.

It's the power that drives, the digital world,
A demonstration of science, with a sharpness that's
curled.
From radios to computers, it's always at work,
A tool that brings us, advancement and perk.

It's the force that enables, the flow of information,
A journey of innovation, with a brightened sensation.
From phones to televisions, it guides our way,
A principle of electronics, that never goes astray.

So let us embrace the Transistor, and all that it brings,
A journey of technology, and a world of wondrous
things.
And in the power of electronics, let us find our way,
To a smarter future, with innovation each day.

Integrated circuits.

Integrated circuits, a marvel of design,
A network of transistors, that works in a line.
A symphony of signals, and logic gates too,
A journey of electronics, that's always in cue.

It's the power that drives, the modern age,
A demonstration of engineering, with a technological
sage.
From laptops to satellites, it's always at work,
A tool that brings us, convenience and perk.

It's the force that integrates, the circuits we need,
A journey of efficiency, with a lightning speed.
From cameras to cars, it guides our way,
A principle of electronics, that never goes astray.

So let us embrace the Integrated Circuit, and all that it
brings,
A journey of technology, and a world of wondrous
things.
And in the power of electronics, let us find our way,
To a smarter future, with innovation each day.

Generators.

Generators, a source of energy so grand,
A machine of motion, that takes us to a different land.
A creator of power, that flows without a cease,
A journey of electricity, that brings us peace.

It's the power that drives, the factories we build,
A demonstration of engineering, with a strength that's still.
From cities to villages, it's always at work,
A tool that brings us, energy and perk.

It's the force that generates, the energy we need,
A journey of reliability, with a steady speed.
From turbines to engines, it guides our way,
A principle of mechanics, that never goes astray.

So let us embrace the Generator, and all that it brings,
A journey of power, and a world of wondrous things.
And in the creation of energy, let us find our way,
To a brighter future, with technology each day.

signals.

Signals, a language of the technological kind,
A message of information, that travels with a bind.
A whisper of sound, a flash of light,
A journey of communication, that shines so bright.

It's the power that drives, the world we know,
A demonstration of science, with a steady flow.
From radios to smartphones, it's always at work,
A tool that brings us, connection and perk.

It's the force that conveys, the messages we send,
A journey of transmission, with no end.
From Wi-Fi to satellites, it guides our way,
A principle of communication, that never goes astray.

So let us embrace the Signals, and all that they bring,
A journey of connection, and a world of wondrous
things.
And in the power of communication, let us find our way,
To a connected future, with technology each day.

Wi-Fi.

Wi-Fi, a network so swift and so strong,
A web of connection, where information belongs.
A sea of signals, that flows through the air,
A journey of wireless, that takes us there.

It's the power that drives, our devices and more,
A demonstration of technology, that opens a door.
From homes to offices, it's always at work,
A tool that brings us, convenience and perk.

It's the force that connects, the world we see,
A journey of accessibility, with liberty.
From laptops to tablets, it guides our way,
A principle of communication, that never goes astray.

So let us embrace the Wi-Fi, and all that it brings,
A journey of freedom, and a world of wondrous things.
And in the power of wireless, let us find our way,
To a connected future, with technology each day.

satellites.

Satellites, floating in the sky so high,
A network of orbiters, touching the sky.
An army of guardians, watching from above,
A symbol of technology, and its boundless love.

It's the power that connects, the world so vast,
A demonstration of science, that outlasts.

From communication to navigation, it's always at work,
A tool that brings us, connection and perk.

It's the force that guides, our travels and ways,
A journey of navigation, that never strays.
From GPS to weather forecasts, it guides us right,
A principle of technology, that shines so bright.

So let us embrace the satellites, and all they bring,
A journey of information, and a world of wondrous things.
And in the power of orbiters, let us find our way,
To a connected future, with technology each day.

GPS.
GPS, a navigation so precise,
A system so simple, yet oh so nice.
A network of satellites, guiding the way,
A tool that brings us, to our desired lay.

It's the power that finds, our location with ease,
A demonstration of technology, that never cease.
From cars to phones, it's always on call,
A tool that brings us, guidance for all.

It's the force that directs, our journey ahead,
A journey of precision, that never goes astray.

From mapping to navigation, it leads us right,
A principle of science, shining so bright.

So let us embrace the GPS, and all it brings,
A journey of convenience, and a world of wondrous
things.
And in the power of navigation, let us find our way,
To a connected future, with technology each day.

Ultrasound.

Ultrasound, a sound beyond the range,
A tool so gentle, yet it brings such change.
A beam of energy, that travels through skin,
A technology so gentle, yet it's power within.

It's the power that sees, what lies beneath,
A demonstration of science, that brings us relief.
From checking on babies, to scanning our heart,
A tool that brings us, a brand new start.

It's the force that reveals, what lies inside,
A journey of discovery, with its gentle guide.
From medical diagnosis, to researching more,
A principle of technology, that helps us explore.

So let us embrace the ultrasound, and all it brings,

A journey of insight, and a world of wondrous things.
And in the power of this gentle beam, let us find our way,
To a brighter future, with technology each day.

Alpha ray.

Alpha rays, a particle so bright,
A journey of discovery, that shines so light.
A stream of energy, that travels through air,
A demonstration of science, beyond compare.

It's the power that illuminates, the path ahead,
A tool that brings us, a new way to tread.
From exploring the universe, to studying the earth,
A tool that brings us, a new sense of worth.

It's the force that uncovers, secrets untold,
A journey of enlightenment, that never grows old.
From mapping the stars, to unlocking the past,
A principle of physics, that will forever last.

So let us embrace the alpha rays, and all they bring,
A journey of knowledge, and a world of wondrous things.
And in the power of this radiant particle, let us find our way,

To a brighter future, with science each day.

Nuclear Fission.

Nuclear fission, the splitting of the atom's core,
Releasing energy, like never seen before.
A force so powerful, it's both a gift and a curse,
Its applications, ranging from power to a war.

It's the source of light, that keeps our cities bright,
A solution for energy, in a world of endless night.
It's the fuel that drives, the machines that do,
A technology that brings us, a brighter tomorrow too.

But beware of its dangers, for it's not all it seems,
It can bring destruction, to countless lives and dreams.
A cautionary tale, of a force we cannot control,
A reminder of science, that must be nurtured, whole.

So let us embrace nuclear fission, with responsibility and care,
A journey of discovery, that requires our watchful glare.
For in the power of this atom, lies both a boon and a threat,
And it's up to us, to use it wisely, with each passing day we met.

Nuclear Fusion.

Nuclear fusion, the fire of the stars,
A force so bright, it reaches near and far.
Bringing light to the universe, energy untold,
A promise of power, that will never grow old.

It's a dance of particles, in perfect harmony,
A reaction that's controlled, by such high energy.
A future of energy, that's safe, clean and bright,
A beacon of hope, that shines with endless might.

But fusion is elusive, a goal that's hard to reach,
A challenge that scientists, have worked to breach.
The quest for fusion, is a journey long and true,
A path to a future, where energy's always anew.

So let us strive, to unlock nuclear fusion's key,
A source of energy, that's bright, safe and free.
For in the heart of this reaction, lies a brighter
tomorrow,
A power that can change, the course of our life's sorrow.

isotopes.

Isotopes, unique elements, each one the same,
With nuclei alike, in their atomic claim.
But different in weight, a distinction small,
Yet impacting the world, in ways we recall.

In nature's symphony, isotopes play their part,
Some are stable, some not, a tale of each heart.
Some light and fleeting, some long and strong,
In the world of the atom, they dance all day long.

In medicine and industry, isotopes have a role,
A tool that's powerful, to heal and to control.
From diagnosing diseases, to powering our homes,
Isotopes are present, in all that we've known.

So let's pay homage, to isotopes rare and bright,
Their contribution, a shining example of might.
For they hold secrets, of the world we call home,
And they continue to change, the way we roam.

MRI.

An image so clear, a glimpse inside our minds,
A machine so wondrous, it's hard to define.
With magnets and radio waves, it gives us a view,
Of the mysteries inside, that we never knew.

The MRI, a tool so powerful and wise,

A window into our bodies, a way to visualize.
With no harmful rays, it opens up the doors,
To our inner selves, and what lies within our cores.

From headaches to tumors, it helps diagnose,
The cause of our pain, and the secrets it hides.
With precision and clarity, it shows us the way,
To a better tomorrow, and a brighter day.

So let's sing its praises, this machine so grand,
A tool that's changed medicine, and our understanding
of the land.
For it gives us insight, into our health and our soul,
And helps us on our journey, to make us whole.

Scintillation.
A spark in the dark, a flash of light,
A discovery waiting, in the dead of night.
The scintillation, a phenomenon so rare,
A glimpse into the mysteries, of the world beyond the
air.

It dances and it twirls, it sparkles and it shines,
A firefly in the night, a symphony of light.
It illuminates our path, and guides us through the night,
A beacon in the darkness, a guiding light.

It reveals what lies beneath, what's hidden from our
sight,

The secrets of the universe, in the darkest of the night.
From particles to gamma rays, it gives us insight,
Into the mysteries of the world, that lie beyond our
sight.

So let us celebrate, this gem of the night,
A symbol of our knowledge, and a source of light.
For it shows us the way, to a world so bright,
And helps us on our journey, towards a new and better
light.

Electric charge.
Electric charge, a force so strange,
A mystery we seek to explain,
With positive and negative range,
It sets the atoms in its chain.

Like opposite poles of a magnet,
Electric charges, they attract,
And when they meet, they will react,
In a spark of light, they can impact.

It flows through wires with ease,
And powers all of our needs,
From lighting up our homes,
To charging phones.

It's what gives electrons their spin,
And makes lightning bolts begin,
Electric charge, a wondrous thing,
It's all around, we just can't see.

In every atom, it's at play,
A force that guides us every day,
Electric charge, so small and bright,
It's the key to our world's light.

Electric field.

Electric field, an invisible force,
A power that holds our world's course.
It's like a wind that pushes and pulls,
Electrons it moves, and energy it fuels.

It's a force so strong, it cannot be seen,
But its effects can be observed, like a king's reign.
It can attract and repel, all in its way,
And bend light too, as it moves and it plays.

From the smallest particle to a massive star,
Electric field holds the universe's bizarre.
It's the spark in a lightning, a flash in a storm,

It's the glue that holds atoms and keeps them warm.

Electric field is a fundamental part,
Of our world and universe, and it plays its part.
It's a dance of positive and negative charges,
A story untold, that never disengages.

Magnetic field.
A force unseen, yet ever so real,
It swirls and dances, a magnetic appeal.
It can push and pull, like a gentle breeze,
Guide the compass needle with such ease.

In the silence of space, it reigns supreme,
A constant presence, an endless dream.
It surrounds the earth, a protective shield,
Defending against cosmic forces that yield.

It's the glue that holds the universe together,
A force of nature, like the changing weather.
It fuels the stars, the birth of light,
The essence of existence, a brilliant sight.

From MRI machines, to generators in power,

The magnetic field holds a vital hour.
It's a fundamental force, a part of life,
A necessary piece, in this grand design.

So here's to the magnetic field, a wonder untold,
A presence that's felt, yet forever so bold.
A force of nature, that will always remain,
Guiding us through life, like a beacon of gain.

Gravitational field.

Gravitational field, a force we can't see
But its impact on us, is constantly
An invisible pull that holds us in place
A vast ocean of power, spreading through space

The Earth and the moon, it holds in its grasp
A dance of celestial bodies, forever clasped
A force so strong, it bends light's path too
And creates a lens that's used to view through

It's a mystery that scientists still study
A force that shapes the universe and our galaxy
From black holes to planets, it's in all things
A force that makes the universe sing

Gravitational field, so vast and so grand
A force that's felt throughout the universe's land
A pull that we can't resist or escape
A force that we can never truly shape.

Capacity.
In every wire and circuit board,
A hidden force we can't afford,
To ignore, this energy stored,
It's the force that keeps us adored.

It waits, like water in a tank,
For its release, with just a crank,
Or a spark, to light up the bank,
Of energy, that's stored in the rank.

And that's what we call capacity,
A measure of how much can be,
Stored, in a battery or a key,
Or a capacitor, for us to see.

In electronics, it's a must,
To know the value of capacity's trust,

To keep the flow of electrons just,
And ensure the performance is robust.

So remember, in every device,
There's a force that holds a great prize,
And that's the capacity, a wise,
Investment for a future sunrise.

Electric current.
Electric current, like a rushing stream,
Flows through the wires, with such power and gleam.
It moves the electrons, a dance so sublime,
Bringing energy to life, with a rhythmic chime.

A force of nature, that's hard to explain,
Electric current, drives the world's train.
From factories to homes, its power we use,
Making life easier, with no need to refuse.

It's not just a flow of electrons we see,
Electric current, also affects you and me.
It powers our devices, keeps lights bright and bright,
A source of energy, throughout the day and night.

So let's take a moment, to appreciate its might,
Electric current, it's the soul of light.
It's the pulse of progress, the future's beating heart,
Electric current, it's a crucial part.

Kirchhoff`s Law.

Kirchhoff's Law, a gift from the past
Two simple rules, that forever will last
The first one says, in a circuit complete
The sum of currents, is always zero, it can't be beat

The second rule, to the conservation of charge
In a closed circuit, must always enlarge
The sum of voltages, around the loop
Is equal to zero, this principle will always hold true

From simple circuits, to complex designs
Kirchhoff's Law, helps us read the signs
Of the flow of energy, and the ways it bends
A powerful tool, that never ends

So let us cherish, this brilliant man's mind
And always use his laws, we shall be kind
To the science of electricity, and how it flows

Thanks to Kirchhoff, and his amazing insights it shows.

Amplifiers.

Amplifiers, so simple yet so grand,
Electric signals they take in hand.
From faint whispers to a mighty roar,
They increase the strength ten times or more.

With their magic touch they boost the sound,
A clearer message they help us found.
In the world of electronics they reign,
Amplifiers, their power will remain.

From radios to phones, TV and more,
Amplifiers play a vital role.
They help us hear what we couldn't hear,
And bring the music, crystal clear.

So next time you hear a sweet refrain,
Or a voice that's clear and free from pain,
Remember the tiny amps at play,
That make the sound come alive today.

Transducers.

A transducer, a silent friend,
Converts energy, never bend.
From light to sound, or force to volts,
Its work is crucial, solves many a faults.

It senses the world, with such grace,
And converts its signals with perfect pace.
A silent messenger, with power untold,
It brings new life, to what once was old.

From machines to nature, it helps us hear,
The whispers of wind, the rustle of deer.
In medical science, its work is bright,
Helping to see, beyond human sight.

So let us pay homage, to this machine,
That transforms the world, in a routine.
It may be small, but its impact is great,
A true hero, in technology's fate.

Electrodes.

Electrodes, conductors of charge,
So vital in many fields and large,
In batteries they store the power,
And also allow us to measure and tower.

In medicine they help us see,
The electrical activity in our body,
In welding they make metal unite,
And in circuits they complete the sight.

From platinum to stainless steel,
There's a wide range of materials to feel,
And each with its own unique feature,
In its application it will have the nature.

In electrolysis they separate the ion,
And in galvanic cells they provide the ion,
In short, the uses of electrodes are vast,
They make our world more technologically fast.

Quantum theory of Light.

Invisible yet shining bright,
Light holds secrets beyond our sight.
It travels far, it travels fast,
Revealing more with every passing cast.

But then it was discovered,
Light behaves like no other.
Quantum theory came to life,
Revealing secrets, ending strife.

Waves and particles, both and one,
Light is both, it has begun.
It travels straight, it travels curve,
In photons, its energy converges.

And with each interaction,
It proves its quantum satisfaction.
Photons leap, they also spin,
Their properties we can't always win.

Invisible and grand,
Light's behaviour we now understand.
Quantum theory of light,
Revealing secrets, shining bright.

Photo-electric effect.
A flash of light, electrons set free,
A phenomenon known as photo-electric.
Einstein first explained with such ease,

A quantum theory that's hard to beat.

With energy packed in each photon,
And electrons waiting at the gate,
When the two collide, like a dawn,
The electrons leap and vacate.

The amount of energy they receive,
Determines their speed and their release,
This effect, a simple belief,
Is crucial in many devices.

From solar cells to TVs,
Photo-electric effect sets them free,
Converting light to electricity,
And bringing us technology.

So next time you see a flash of light,
Think of the electrons taking flight,
And know that this wondrous sight,
Is the magic of photo-electric effect tonight?

Compton scattering.
Compton scattering, a fascinating sight
The interaction of photons, with all its might
The energy exchange, a fascinating play
Of photons and electrons, in a cosmic ballet dance way

The photons collide, with electrons so free
And change their path, as they scatter and flee
The energy lost, to electrons so small
A shift in the wavelength, and a rise in the hall

The theory of Compton, so elegant and true
Explains the behaviour, of light so anew
A key part of quantum mechanics it be
In understanding light, its interaction and energy

So next time you see, light scattered around
Think of Compton scattering, and its impact profound
In shaping our understanding, of this cosmic light
That illuminates our world, both day and night.

De-Broglie wavelength.

The smallest things in our world,
Can hold secrets that swirl,
A wavelength is their key,
De-Broglie's theory, for you and me.

Atoms and electrons, so small,
With movements that astound us all,

The wavelength they possess,
Is part of science, none can guess.

It's a connection, between mass and wave,
That helps us to observe and probe,
The mysteries of the universe,
With De-Broglie's wavelength as a verse.

So as we look at all around,
We can now understand the sound,
Of the smallest things in our sight,
Thanks to De-Broglie's shining light.

Davisson-Germer experiment.
The Davisson-Germer experiment, a tale untold
Of electrons, waves and atoms, growing bold
It was an unexpected discovery they found
A result that caused a revolution 'round

A beam of electrons, scattered in space
Its path observed with a diffraction trace
And what they saw was truly a sight
A pattern of waves, bringing forth light

The electrons' path was not so clear
A wave-like motion, causing great cheer
A confirmation of de-Broglie's theory
A bridge between waves and particles in glory

This experiment changed the world we knew
A glimpse into the quantum realm, all true
It showed us how the universe behaves
A journey through matter and beyond the waves

So let us remember the Davisson-Germer feat
A landmark in science, that can't be beat
Its legacy lives on, guiding us still
Towards a greater understanding, as we will.

Group velocities.
Group velocities, they move as one
In waves they travel, like a rising sun
Together they flow, from place to place
In harmony they race, with a steady pace

Each one unique, yet intertwined
Their paths aligned, like a gentle wind
They work together, to reach their goal

In perfect balance, like a scale so whole

Their speed and direction, a mystery to see
A complex interplay, between energy and frequency
But scientists have found, this secret to reveal
The group velocity, can help them understand the appeal

So when you hear, the sound of waves
Remember these velocities, that nature saved
Their journey untold, a story untold
Their movement bold, with a heart so bold.

Phase velocities.

Phase velocities, through waves they roam,
Carrying signals, like a messenger home.
Moving with speed, through solids, liquids, gas,
Traveling with care, they carry the task.

Invisible forces, that make up our world,
Vibrations and motions, their story is told.
From sound to light, their journey begins,
Carrying data, where it needs to win.

In the depths of waves, the phase velocity lies,
Its pace sets the stage, for signals to arise.
It's a measure of speed, that travels with time,
The driving force, that makes waves align.

In materials, it varies with frequency,
Higher the frequency, faster it moves with glee.
In air or vacuum, it's the speed of light,
In solids, it slows down, but still shining bright.

From communication, to medical technology,
Phase velocities, play a vital role in society.
They carry information, with speed and grace,
A phenomenon of physics, time and space.

Wave amplitude.
The rise and fall, a rhythmic beat
Of crests and troughs, with grace and heat
A pulse that travels through the air
The force of nature, everywhere

This is the amplitude, the wave's might
That tells us of its power, its sight
It moves with purpose, with will to please
Bringing us joy, and beauty to seize

It dances in the ocean's swell
And shimmers in the light that fell
It carries with it, sound and song
And brings the world alive, all day long

So let us cherish this simple thing
This gift of nature, a beautiful being
For in its waves, we find our peace
And all the wonder, that will never cease.

wave functions.

Waves in motion, such a sight
With every crest and every trough, what a sight
But what makes them move, with such delight?
The answer lies within, the wave function's might

It's a mathematical tool, for us to comprehend
The behaviour of waves, their form and extend
From light to sound, from the sea to the land
The wave function describes, its every strand

It's the blueprint, for waves in the sky
The equation that predicts, their shape on high
With every crest and trough, it tells us why

And how they move, with grace and glide

So if you want to know, what makes waves dance
Take a closer look, at the wave function's chance
And marvel at its power, and the waves' stance
For it holds the key, to their rhythm and trance.

gamma ray microscope thought experiment.
Gamma rays, a piercing light
Invisible to human sight
With energy beyond compare
The microscope we dare to dare

To study matter at its core
We shine this ray and so much more
The atoms dance, the electrons spin
In this thought experiment, we win

The insight we gain, beyond belief
Into the workings of relief
Invisible processes we can see
And learn so much, it's all so free

A dance of particles in space
The gamma ray gives us this grace

To explore what lies inside
And see the beauty we can't hide

So let us shine this brilliant ray
And learn the secrets in its way
The universe, so vast and wide
We're just scratching its surface, but it's a beautiful ride.

Heisenberg uncertainty principle.

The quantum world's a strange and curious place
Where things can be both a particle and space
It's ruled by the Heisenberg uncertainty principle
That tells us some things we can never relevantly know
sequence

It states that the more we know a particle's place
The less we can know of its speed and its pace
And vice versa, a strange and mind-boggling state
One that can only be grasped by those who are great

It's a fundamental aspect of quantum physics
And affects all the matter and energy we see
It limits the precision of our measurements
And provides an eerie glimpse of reality

It challenges our classical understanding of space
And how we observe the world with our eyes
But it helps us understand the strange world below
The one that's ruled by probabilities and lies

So embrace the uncertainty principle with pride
And let it challenge the way you perceive
The strange and mysterious world of the subatomic
And all the wonders that science can achieve.

Canonical pair of variables.
In the world of physics, a pair so grand,
With variables that always go hand in hand.
One reveals position, the other its speed,
Together they form the most perfect indeed.

This duo, known as the canonical pair,
Defines the essence of quantum affairs.
Their uncertainty, Heisenberg did reveal,
A principle that will always appeal.

They dance and they twirl, these two partners divine,
Each complementing the other, like sun and sunshine.
Their values forever intertwined,

In a relationship that's truly entwined.

And so, in this universe so vast and so bright,
This canonical pair leads the way with its light.
Guiding us to a deeper understanding,
Of the mysteries of quantum commanding.

ionization potential.
In the world of atoms and electrons,
There's a potential waiting to be unlocked,
A measure of energy, so fine and precise,
That sets the stage for chemical bonds to unite.

It's the ionization potential, a vital force,
That separates electrons from their host,
A quantum leap that alters their course,
And launches them into the great unknown.

With each jump, the atom is transformed,
Its energy level altered, never conformed,
To the laws of classical physics, reformed,
As it enters a new realm, beyond.

So if you seek to understand the world,

And the mysteries that lie within,
Know that the ionization potential swirls,
At the heart of the atomic spin.

Schrodinger equation.
In the heart of physics lies a truth untold
A mathematical formula that's worth its gold
It's called the Schrodinger equation, a key
To unlock the secrets of matter and energy

It shows the wave nature of matter, so grand
The probabilities of finding it, in a certain land
And how it evolves over time, that's the thrill
Explaining everything from the smallest quill

It's a key to unlock the secrets of the atom
It's a rule that helps us to understand the quantum
It's a mystery, wrapped in a paradox, that's true
The Schrodinger equation, that's how we know what's true.

With it, we delve into the unknown depths of space
It's a bridge that connects the quantum to our place
And thus it stands, this powerful equation

An emblem of knowledge, a path to revelation.

probabilities and normalization of wave.
The universe's behaviour, it's quite complex and grand,
We've tried to understand, with theories and math
that's sound,
One such concept that can't be ignored,
Is the idea of probability, for which we've explored?

The waves that describe the particles we see,
Are not fixed and definite, for all to believe,
They hold potential and chances, we do confess,
To be here, or there, in multiple locations at once, we
guess.

But we need to make sure, that these waves don't grow
wild,
That the sum of all probabilities, is kept under control,
and mild,
For the sake of accuracy and the science we hold dear,
We apply the rule of normalization, to keep the waves
near.

So if you delve deep, into the mysteries of life,

Just remember, the probability and normalization of the wave,
Is a crucial part, of the Schrodinger equation we rave,
Helping us understand, the microscopic world with strife.

probability current densities wave.
A current of chance that flows through space,
With every movement, it finds its place.
A density of waves, both high and low,
A measure of where particles might go.

It's not a force that we can see,
But like a river, it's always free.
To move and flow and shift its shape,
The essence of quantum's grand escape.

It's a delicate dance of particles and waves,
A balance of potential and probabilities weaves.
A symphony of chance and uncertainty,
That leads us to the quantum reality.

It tells us where the particle might be,
And helps us understand its destiny.

A key to unlock the mysteries of life,
The probability current density of quantum's rife.

Millikan oil drop experiment.

In a lab with a flask so grand,
Millikan began his master plan,
To measure electrons with such ease,
And find their charge, to say with pleas.

He took an oil drop, light and small,
And placed it in an electric hall,
Where charged plates, both high and low,
Would push and pull, to make it go.

With voltage and light, he found the way,
To measure charge, both night and day,
And when he measured, oh so well,
He found that electrons charged by shell.

His experiment, still well-known,
Helped us understand what's grown,
And now we see, with clarity,
The smallest of charges, with integrity.

So let us all, with Millikan's might,
Remember this experiment tonight,
And always strive, to measure more,
Of what makes the world, both light and core.

energy eigenvalues and Eigen functions.

Energy eigenvalues, Eigen functions,
Solutions to Schrödinger's equation,
A quantum world, where precision leads,
To predictions, with such detail and reason.

A wave with a form, so unique,
Characteristics of its own, to seek,
The eigenvalue, a constant divine,
Of energy, that stays aligned.

In a world, where probabilistic views,
Reign supreme, and all that's true,
The Eigen function, a way to find,
The distribution, of what's confined.

From particle in a box, to hydrogen's states,
Energy levels, of atoms and their weights,
The eigenvalue and Eigen function, unite,

To bring to light, the mysteries of the night.

So here's to the concept, elegant and grand,
A cornerstone, of quantum mechanics' band,
Energy eigenvalues, and Eigen functions, a key,
To unlocking the secrets, of the quantum sea.

particle in a box.
In a box so dark and deep,
A particle does gently sleep.
It moves about with boundless grace,
In a space it can't escape.

Its energy, confined and still,
Is held within the box until
It finds a way to be set free,
And escape its captivity.

But until that day arrives,
The particle just peacefully thrives
Within its box, a curious sight,
A mystery waiting to ignite.

Its motion is both strange and true,

A wave and particle in view.
The box is small, yet grand its stage,
A perfect place for it to engage

In quantum dance, its energy
A string of numbers, simple, free.
A tale of physics and of art,
A particle in a box, a work of heart.

Quantum dot.
A tiny world of electrons bound
A Quantum dot, a secret found
So small, yet full of might
Emitting light, with colours bright

Electrons dance, a cosmic show
In a space so small, you'd never know
Quantum rules, they do apply
With wave-like behaviour, they defy

The energy levels, they define
And trap light, a work of art, combined
The future holds, a promising sight
With Quantum dots, bringing new light

In technology, they play a part
With applications, straight from the heart
From solar cells, to quantum gates
Quantum dots, with endless fate

So small and yet, with power untold
A Quantum dot, a treasure, unfold.

Quantum mechanical scattering.
A photon in the void, like a tiny seed,
Traveling through space with incredible speed.
It meets a particle, small and confined,
A collision awaits, the future designed.

The outcome is known, with certainty and grace,
Quantum mechanics sets the time and place.
The photons depart, in a different direction,
With its path altered, by a small reflection.

The particle scatters, like the stars in night,
Its trajectory altered by a single light.
This dance of particles, so finely choreographed,
Is guided by nature, its laws never debated.

The future is set, with probabilities so high,
A wave of potential, with amplitudes that defy.
The path of the photon, forever changed,
In the world of quantum, so strange.

So let us embrace, this mysterious fray,
Where particles collide, in a unique display.
For in the quantum realm, where matter and light unite,
The future is uncertain, yet certain at sight.

Quantum tunnelling.
Quantum tunnelling, a mysterious dance,
A particle penetrates a chance.
Through barriers, solid walls so tough,
It slips and slides, but does enough.

With no energy to clear the way,
It moves beyond, each night and day.
A surreal phenomenon, beyond belief,
It opens doors and finds release.

So tiny and yet so full of grace,
This tunnelling particle, a mysterious place.

It takes us to a world so new,
With wonders waiting, a quantum view.

It shows us how the universe behaves,
With particles that penetrate and caves.
The world is not what it may seem,
Quantum tunnelling is a science dream.

nuclear force.
A force so strong, it holds the nucleus tight,
Binding protons and neutrons, day and night.
It's a force that cannot be seen or touched,
But it plays a critical role, that can't be denied or
overmuch.

It's called the nuclear force, and it's a mystery,
One that scientists have worked on, so meticulously.
It's not like electromagnetic or gravitational pull,
It's a force that's unique, and breaks every rule.

It's the force that determines the size of the nucleus,
It's the force that controls the stability and the fuss.
It's the force that keeps the protons from repelling
away,

And creates the nucleus, that holds our atoms in play.

The nuclear force is the glue that holds the nucleus
together,
It's the force that shapes the nucleus, in any weather.
It's a strange and wonderful force, that's still not
understood,
But one that plays an important role, in the world of the
good.

Liquid Drop model.
A droplet of liquid, small and round,
With secrets untold, waiting to be found.
The nucleons inside, with forces they bind,
Making it strong, an atom combined.

A delicate balance, between protons and neutrons,
A force so strong, it's like magic potions.
It holds the nucleus together, in a tight embrace,
Withstanding the forces that might break the space.

The model of liquid, a metaphor we see,
For the forces within the nucleus, oh so free.
It helps us understand, the secrets untold,

Of the forces that keep the nucleus bold.

So next time you see a droplet of rain,
Think of the secrets it holds, not in vain.
For it's a model of nature, so precise,
That tells us the story of the nucleus in disguise.

binding energy.
Binding energy, a force so strong
That holds particles where they belong
Nuclei, it keeps them in a knot
A bond that's tough, that cannot be fought

It's measured in units, quite precise
A number that gives insight and advises
Of a nucleus' strength and its state
And how tightly its protons and neutrons mate

This energy, hidden but so grand
Is released when a nucleus expands
In nuclear reactions, it's set free
And provides energy, for you and me

It's a powerful force, binding energy

That helps us understand nuclear chemistry
A crucial component of our world
A concept that forever will be unfurled.

Nuclear Shell Model.

Nuclear Shell Model, a tale untold,
Of protons and neutrons, in nuclei bold,
Orbitals within, a perfect design,
Energy levels that seem to align.

Nucleons inside, with spaces to fill,
Following a pattern, their roles to fulfil,
Spherical in shape, with symmetry so true,
Magic numbers that give rise to stability too.

Electronic shells, much like in atoms,
Nucleonic shells, in nuclei that are calm,
Define the structure, of atomic nuclei,
Energy levels, that keep it from debris.

A tale of physics, a tale of might,
The Nuclear Shell Model, its secrets in sight,
Of forces inside, and stability so great,
It's a model, that we all should celebrate.

magic numbers in nuclear.

Nuclei so small, and yet so grand,
With energy that's both binding and bland,
A model it has, a shell that's divine,
Making sense of the magic numbers that align.

Each number unique, with purpose and place,
Gives stability to the atomic race,
With protons and neutrons all in a row,
The magic numbers, a mystery now known.

Eight, twenty, fifty-two,
The numbers that nuclear forces imbue,
A harmony that's not hard to explain,
Making the nucleus a stable domain.

The nucleons dance, with balance and grace,
Defying the odds, they take their place,
In a symphony of energy and might,
That's both beautiful and sheer delight.

So if you seek to understand,
The magic of nuclei and the nuclear band,
Look to the shell model, a tale so grand,
Of energy and forces, in a beautiful land.

radioactive decay Mean life and half-life.

The atom is not as stable as one may believe,
With change happening, in each molecule and heave.
The nucleus may decay, with time passing by,
Emitting particles and energy, that reach the sky.

Mean life and half-life, are terms we must define,
To understand the speed at which decay occurs, it's a
sign.
Half-life, the time it takes, for half the substance to go,
Mean life, the average, before it starts to show.

Radioactive decay, it's not a static fate,
It varies from substance to substance, it's not too late.
It's important to measure, this process in time,
To understand the dangers, and keep people safe, it's a
rhyme.

Pauli's prediction of neutrino.

Oh Pauli, a mind so bright,
With a theory that shone so bright,
Predicted the presence of a particle,
One so small, its existence was doubtful.

A neutrino, a ghostly soul,
So elusive, it's never quite whole,
With no charge, or mass, it seems,
It travels through all matter, with its own beam.

From the sun, to the depths of space,
It flows through, with an effortless grace,
It's interaction with matter so rare,
It took time to prove Pauli right, with care.

So here's to Pauli, and his foresight,
And the particle that he brought to light,
The neutrino, so mysterious, yet real,
A reminder of how much is yet to reveal.

energy-momentum conservation.
In the world of physics, we find,
A principle of conservation kind.
Energy and momentum must stay,

Together in their balance, come what may.

A single law that rules them all,
In every reaction, great and small.
No matter what the process may be,
This principle shall forever be.

From colliding particles with might,
To stars exploding in the night,
This rule holds true in every case,
Conserving energy and pace.

So let us hold this truth close at heart,
And never let this principle depart.
For in the universe, great and vast,
Energy-momentum conservation shall last.

Metastable states laser.
Metastable states, like stars so bright,
Glimmer in the laser's beam of light.
A resting place, they seem so still,
But energy waits, with just a thrill.

A trigger comes, a spark so slight,

And they ignite, in glory bright.
A sudden flash, a burst of power,
Releasing energy in every hour.

The laser shines, a ray so pure,
With energy that's both bright and sure.
And from the metastable state so high,
Comes a burst, that lights the sky.

So hold on tight, for this is just the start,
Of a journey, that sets the world apart.
For metastable states, they hold the key,
To unlock the power of laser energy.

Spontaneous emission.
In the world of photons and atoms so bright,
A secret was hiding in plain sight.
It's called spontaneous emission, a photon in flight,
From an atom that's excited and shining so bright.

The energy levels in atoms, are not quite so fixed,
And in their excited state, they can mix and mix.
When this happens, a photon is released, with delight,
And in a flash of light, the atom returns to its sight.

This phenomenon, so beautiful and free,
Is responsible for the stars and you and me.
For the light that guides us, and the world we see,

Is the result of this magical, spontaneous emission?

So next time you look up at the sky so blue,
Think of this process, and how it all came through.
And remember that even in the smallest thing,
There's a beauty and magic that makes our hearts sing!

Stimulated emissions.
In the heart of a laser, light starts to grow,
With atoms that emit, a brilliant light show.
A photon arrives, to stimulate the scene,
And suddenly more, photons are seen.

The light builds up, in a chain reaction,
The photons come fast, with no inhibition.
The atoms in sync, like a marching band,
Each one in tune, with the photons at hand.

Stimulated emissions, the key to this dance,
A beautiful light show, a true chance.
For the light to be bright, and in perfect harmony,
It all starts with this, scientific synergy.

So next time you see, a laser in action,

Remember the physics, and the magic that's woven.
With stimulated emissions, at the heart of it all,
The light shines bright, and stands tall.

Optical Pumping laser.
A beam of light, a ray so bright,
A pump to boost, with all its might,
The atoms inside, with energy to spare,
With photons bright, beyond compare.

The photons dance, a dazzling show,
With energy and power, making it grow,
The electrons rise, from low to high,
And lasers are born, up in the sky.

A beam so pure, a line so fine,
With photons aligned, in a single line,
A beam of light, so intense and bright,
A tool of science, with endless might.

So if you want to light up the night,
And make the world a brighter sight,
Use optical pumping, with all its might,
And watch the lasers, shining so bright.

Thermodynamics.

In the realm of thermodynamics,
Where energy reigns supreme,
The laws of science are born,
And studied by the curious team.

Heat, work, and internal energy,
Are concepts we must all know,
Entropy and temperature,
Their relationship will show.

From the first law, we see
That energy cannot be destroyed,
But it can change its form,
As it moves through space and time.

The second law, a guide,
Of entropy's increasing trend,
Shows how systems will change,
Towards a maximum end.

The third law states
That absolute zero is our goal,

But it remains unattainable,
An ideal that takes its toll.

So, in the world of thermodynamics,
We learn and understand,
How energy behaves,
And how it shapes our land.

heat.
Heat, the energy that flows,
From high to low, it goes,
The source, the fire, the spark,
That keeps us warm in winter dark.

The heat that rises in the air,
That warms the land, it's everywhere,
The light of sun, the force of fire,
The heat that makes the earth transpire.

It's in the water, in the land,
It's in the movement of our hand,
It's in the friction, in the grind,
The heat that gives our world a mind.

The heat that drives the engine's steam,
That boils the water in the team,
The heat that cooks our food and warms,
The heat that gives life to our forms.

The heat that dances in the flame,
The heat that drives our weather game,
The heat that brings the thunder clap,
The heat that makes the world come back.

Heat, the energy that we need,
It's in the air, it's in the seed,
The source of life, the light of fire,
Heat, it's what makes our world entire.

temperature.
Temperature, what a peculiar notion,
A measure of heat's tumultuous motion.
It rises with the sun and falls with night,
A gauge of all that's warm and cool, what's right.

It's not a substance, no, not a thing,
But a concept that helps us understand,
How energy flows, what makes it shift,

In solids, liquids, gasses and life.

From boiling water to frozen ice,
Temperature speaks of change and price,
Of states and processes, so bold and bright,
That shape our world, from day to night.

So let us use this tool with care,
And cherish all that it can share,
For temperature, it tells us so much,
About the world and its underlying touch.

black holes.
In the darkness of space so vast,
Lies a place where light cannot pass,
A void so deep, a pit so black,
Where all matter collapses to pack.

It's a monster of gravity untold,
Where time and space both take a fold,
Not even light can escape its pull,
And all that enters is just gone, null.

It's a mystery, a paradox, a riddle,

A place where science can't seem to muddle,
Its properties, its behaviour, its fate,
All left for us to ponder and debate.

But despite its ominous air,
It's a wonder that we can't compare,
A marvel of the universe's might,
A black hole, a cosmic sight.

Hamiltonian.
A force of nature, a force to be reckoned,
With power to shape, a universe beckoned,
The Hamiltonian holds, a key to the stars,
In a realm of physics, where time and space bars.

It's a mathematical beast, tamed by the wise,
Describing a system, with equations precise,
It calculates energy, a universe's fate,
From the smallest quanta, to a celestial state.

It holds secrets, of matter's behaviour,
With motion and energy, in perfect flavour,
From mechanics to quantum, it's a universal tool,
In a quest for knowledge, it's a guiding jewel.

The Hamiltonian's power, never to be ignored,
It's the foundation, for theories to be stored,
In a universe of wonder, it holds the key,
To unravel mysteries, for all to see.

vacuum pump.
A vacuum pump, a tool of might,
With power to reach new heights.
It creates a void so pure,
A place where air no longer cures.

In this place of emptiness,
It proves its true worthiness.
For research and production, it's a must,
A vacuum pump's power is a trust.

It works to remove gas and air,
A process so precise and fair.
With every molecule it extracts,
The vacuum it creates, exacts.

It's used in many industries,
From electronics to sciences.

For studies, it's a valuable tool,
A vacuum pump, a wonder so cool.

So next time you think of a vacuum pump,
Remember the power that it holds,
In its ability to create the void,
A power, forever to be told.

Boyle's Law.

Boyle's Law, a wondrous sight,
A gas's behaviour in plain sight.
It states that pressure and volume unite,
In an inverse relationship, absolute.

A gas confined in a space,
Will change with pressure and its pace.
Increase the pressure, volume will decrease,
And vice versa, this law won't cease.

The man who discovered this so bright,
Robert Boyle, with his insight,
His experiments and tests so right,
Brought us this law, a gift of might.

This law we use in daily life,
From weather forecasts, to dive,
Into the depths of ocean strife,
Boyle's law is with us, guiding light.

So let us celebrate this day,
And learn more, in every way,
About this law, that makes us say,
Boyle's Law, a discovery that'll stay!

Carnot engine.

A machine of magic, so efficient and rare,
It drives the forces that make the world fair,
No friction, no loss, a perfect machine,
The Carnot engine is the ultimate dream.

It takes heat from a source, and turns it to work,
A cycle so simple, its rules never smirk,
It takes from the hot, and gives to the cold,
And in this transfer, it's story is told.

The efficiency of this engine is grand,
It's limited only by the temperature at hand,
The difference between hot and cold must be wide,

For the Carnot engine to work with pride.

So with every turn of its wheels and its gears,
It reminds us of science, and the wonders it bears,
For it's a symbol of progress, a shining light,
The Carnot engine, a symbol of might.

Theory of the Motive Power of Heat.

In the world of thermodynamics,
A theory was put forth with precision,
By Sadi Carnot with insight so rare,
Of the motive power of heat, a vision to share.

He said, "Heat flows from hot to cold,"
A principle simple, yet bold,
And with this idea, he paved the way,
For a heat engine's workings, to understand and say.

Carnot's engine, a masterpiece of its kind,
With no friction, no losses of any kind,
Efficiency high, a marvel to see,
It set the standard for thermodynamics, you and me.

The theory still stands, unshaken with time,

Guiding engineers, technicians, and scientists alike, to climb,
Higher and higher, in their quest for power,
With the motive power of heat, at the core hour after hour.

Laws of thermodynamics.
The Laws of Thermodynamics,
A set of principles we've learned,
That govern energy and heat,
Their actions can't be out turned.

The first law speaks of energy,
Conservation is its name,
It tells us that energy,
Remains constant in the game.

The second law brings us entropy,
An increase, so we're told,
A measure of disorder,
That grows as heat grows cold.

The third law's one of entropy,
It speaks of absolute zero,

A state in which entropy,
Would cease to grow and glow.

These laws, they shape our world,
From the biggest stars to littlest seed,
And understanding them, is crucial,
For our knowledge, our future, we need.

Zeroth law of thermodynamics.
Zeroth law of thermodynamics
A rule that's foundational, oh so true
States that systems, in thermal contact too
Will reach thermal equilibria, with ease
When temperatures are equal, at least

It lays the foundation, for all to know
Of heat and temperature, and how they flow
In physics and engineering, it holds a key
To understanding, the mechanics of energy

With the zeroth law, as our guiding light
We can study, and understand with might
The behaviour of systems, in thermal space
And make predictions, with a steady pace

So let us cherish, this fundamental law
That brings order, to a world in awe
Of the mysteries, of heat and its flow
And use its teachings, to make things grow.

Isobaric process.

An isobaric process, a thermodynamic route,
Where pressure remains constant, without a doubt.
Heat is exchanged, but the force stays the same,
And the volume changes, it's not an ordinary game.

The state of a gas, determined by two,
Pressure and volume, they guide what we do.
In an isobaric process, one stays put,
While the other changes, and energy is cut.

Adiabatic or isothermal, it's a choice to make,
But isobaric process, is the one to take,
When pressure must remain, and volume adjust,
It's a delicate balance, of energy and trust.

From a gas in a piston, to a balloon in air,
An isobaric process, is always there,

Heat is added or removed, with care,
And the pressure stays constant, everywhere.

So next time you think, about thermodynamics,
Remember the isobaric process, it's not a sin,
It's a way to describe, the flow of heat,
And understand the balance, of pressure and heat.

Isochoric process.

An Isochoric process, with constant volume so rare,
A path of thermodynamics, no change in size to bear.
Heat transfer, work done, internal energy too,
This process unchanging, with physics true.

In a closed system, no change in the space,
A state of constant volume, a fixed pace.
Temperature, pressure, internal energy rise,
A cycle of changes, a thermal surprise.

A process of simplicity, with naught to complicate,
In a closed system, with volume to dictate.
Thermodynamics' essence, a fundamental rule,
An Isochoric process, so simple and cool.

Isothermal process.

Isothermal, a process so fair,
Where temperature stays constant and rare.
A perfect example of thermodynamics,
A perfect balance of heat and mechanics.

Heat flows in and out with care,
The temperature stays steady and fair.
A constant temperature is what we see,
As the heat flows in perfect symmetry.

From gases to liquids, all is well,
In this process, thermodynamics excel.
No matter what, temperature stays the same,
In a state of balance, where all is to gain.

In physics and engineering, it's a name,
A process to remember and claim.
Isothermal, a perfect law of heat,
A study of thermodynamics, where all is neat.

Adiabatic process.
An Adiabatic process, a journey it takes,
Heat exchange it makes, with surroundings it forsakes.
Thermal contact it shuns, energy it conserves,
Transformation it runs, through phases it observes.

With pressure and volume, it goes up and down,
A relationship that's renowned, in thermodynamics town.
The work done in this phase, depends on initial and final state,
It's an ideal way, for heat and energy to debate.

From gases to solids, from liquids to plasma,
This process carries, the thermal signature drama.
In engines and turbines, it plays a vital role,
In conversion of heat, to mechanical goal.

In the Adiabatic journey, entropy remains the same,
Heat transfer is limited, by an impermeable membrane.
It's a way of nature, to show us its path,
From thermal energy, to work that's mass.

Isentropic process.

Isentropic, a process that's rare
No change in entropy, everywhere
The volume and pressure, they do shift
But entropy remains, it's a gift

From a state of gas to a state of liquid
Or from solid to gas, oh so vivid
The entropy remains constant and true
In this process, the change is few

It's like a journey through a steady climb
With no pit stops, it flows in time
It's the most efficient, that's for sure
Conserving energy, it's the cure

In power plants, it's a common sight
Maximizing efficiency, it brings delight
From turbine to condenser, it's the way
In isentropic process, energy's saved
So if you seek a process, efficient and grand
Isentropic, it's the one at hand.

Isenthalpic process

The pressure and enthalpy balance,
In an isenthalpic process quite grand,
No heat added, no heat released,
A path so constant, it won't cease.

In this system, the enthalpy stays the same,
While the pressure may fluctuate or change,
A thermodynamic curve that's unique,
With properties both simple and sleek.

This process is used in many fields,
From power plants to pumping yields,
It brings us a deeper understanding,
Of how energy flows, expanding.

So let us remember this process fair,
Isenthalpic, with stability rare,
A key player in thermodynamics,
A journey that's simple, yet full of magic.

Conjugate variables in thermodynamics

In thermodynamics, there's a pair we must see
Conjugate variables that, in balance, agree

One, work done, and the other, heat flow
Together, they show how energy can grow

In an engine, work's done as the gas expands
Heat flows out, and energy demand
And as the process moves from state to state
These two variables dictate its fate

In an isothermal change, temperature's constant
Heat in and out, their difference apparent
In adiabatic, work is done with no heat flow
Isentropic, entropy stays, energy goes

These pairs, conjugate, in thermodynamics play
A dance of energy, in a balance they sway
Each process, each change, they describe with ease
The flow of energy, and its interplay with degrees.

Boltzmann distribution law

In a sea of atoms, they move and they sway,
Each with energy, on its own little way.
The laws of nature, so brilliant and bright,
Can help us describe this chaotic delight.

Boltzmann's law, a master of kinetics,
Tells us how particles move and distribute.
With each particle's energy in view,
It helps us understand, what it wants to do.

When we look at a gas, it seems quite chaotic,
But with this law, we can be quite fanatic,
It tells us how many are low and high,
And predicts their behaviour, with a watchful eye.

The formula's simple, but the meaning's grand,
It shows us how energy is distributed in the land,
And how as the temperature rises, they go,
To higher and higher energy levels, with a flow.

So if you want to understand, how particles behave,
Boltzmann's law, is the way to be brave,
It shows you how energy and matter combine,
In a world of motion, that's truly divine.

Gibbs paradox

Gibbs paradox, a twist in thermodynamics,
A concept that puzzles and confounds us,
The entropy of a mixture, its energy so grand,

Seems to defy the laws of this physics land.

We mix two gases, with temperatures so bright,
But the entropy of the mixture doesn't seem quite right,
It's greater than the sum of its parts, we see,
A mystery that's eluded physicists, like you and me.

The paradox lingers, its cause not yet found,
But scientists and researchers, their search is un-bound,
In the world of thermodynamics, this conundrum reigns,
An enigma that will keep us searching, till it gains.

The laws of thermodynamics, so strong and so true,
But Gibbs paradox, its presence we cannot subdue,
It challenges us to delve deeper, to understand,
The secrets of this world, with the knowledge of its
hand.

So let's embrace the paradox, its lessons to impart,
To further our understanding, and fill our knowledge
chart,
For in the quest for answers, we grow and we mature,
And in the end, the truth, it will surely procure.

Time dilation

Time dilation, a strange effect,
Of Einstein's theory, we expect.
With speed and gravity, it's combined,
Making time seem, differently aligned.

A clock in motion, ticks slow,
Compared to one, standing still, you know.
The faster you travel, the more you'll see,
Time dilation, a mystery.

A spacecraft flying, near speed of light,
Returning, to find, less time in sight.
From Earth, it ages, just a bit,
But to its crew, it feels like a full kit.

It's a fascinating, bizarre idea,
That time is not absolute, as it appears.
It changes, with speed and gravity's hold,
Leaving us with, stories untold.

So next time you see, a racing clock,
Think of time dilation, and have a mock,
For this strange effect, is all around,
In the world of physics, it's profound.

length contraction

Length contraction, a peculiar dance,
In the world of physics, it takes a chance,
As velocity grows and speed takes flight,
Lengths change and objects shrink in sight.

Once you thought that space was absolute,
A constant that could never refute,
But Einstein's theory, with grace and ease,
Explained how objects in motion, they decrease.

In a rocket ship, as you race through space,
Time dilation and length, a curious chase,
The closer to light, the slower time goes,
And the shorter the length, as the story goes.

Relativity rules, with a twist and a turn,
In a world where mass and energy interchange,
Length contraction, a strange and unique part,
In the physics of our universe, a brilliant art.

Equivalence of mass and energy, $E = mc^2$

The energy, it's a fascinating thing
In its core, it holds the power of a king
Its motion through space, its potential to create
But with matter, it's intertwined, in a bond that's so great

And so it was Einstein who gave us the key
To unlock the secrets of this energy
He showed us that matter and energy unite
And from that, he derived the famous sight

E equals MC squared, a simple equation
Yet one that holds a truth of great sensation
It shows us that matter can be converted to light
And that energy holds the power of might

This concept has revolutionized our view
Of how the universe functions, so true
For now, we know that energy's never gone
It just changes form, a concept now dawned

So hold tight to this idea, and never forget
That energy and matter, they are not to be met
For they are one and the same, a fact of the cosmos
And with this knowledge, we can reach new heights, beyond us.

theory of relativity

A theory of the universe, a mind-bending feat
One man named Einstein, his ideas hard to beat
The fabric of space-time, he saw it in a new light
A structure so flexible, it bends by mass's might

Gravity, no longer a force that pulls down
But a warping of space-time, an object falls to the ground
Time and space, no longer fixed and divine
Intertwined and relative, to each observer's line

Mass and energy, intertwined in a grand equation
$E=mc^2$, the famous law of our creation
As energy increases, mass will always grow
In this new worldview, the universe is on the show

From black holes to stars, to photons in flight
The theory of relativity, helps us see things right
A universe of wonder, and mysteries untold
Thanks to Einstein, our minds have grown bold.

Quantum entanglement

Quantum entanglement is a phenomenon so strange,
A connection so strong, it defies exchange,
Two particles can be intertwined,
Their states are correlated, but never aligned.

A twist in the fabric of space and of time,
A bond so mysterious, it's hard to define,
But still it remains, a truth that's so clear,
A principle of quantum physics we hold so dear.

It's like magic, yet it's not just a trick,
It's a phenomenon that's real, it's no gimmick,
Two particles, apart, but always linked,
A connection so strong, it cannot be blinked.

The distance may grow, and space may expand,
But still the bond remains, forever unspanned,
Quantum entanglement, a marvel to see,
A connection that defies reality.

quantum harmonic oscillator

In the world of quantum mechanics,

Where particles can be both waves and specs,
There's a concept that's well known,
The quantum harmonic oscillator, it's grown.

With its energy levels that can be counted,
And states that can be quantized, it's mounted,
A foundation of quantum theory,
Its principles used widely in physics and engineering,
you'll see.

With a particle bouncing back and forth,
In a well that's bound from north to south,
The oscillator is a simple system,
That's been studied and analysed, it's been given the
primacy.

It's a model of a single degree of freedom,
That's fundamental to our understanding, that's the
notion,
Of quantum mechanics, its predictions are sound,
It's helped us to build a better world all around.

So, the quantum harmonic oscillator is a beautiful sight,
With its simple and elegant principles, it brings delight,
To all who study it, and gain insight,
Into the fascinating world of quantum mechanics, it's
just right!

quantum chromodynamics.

In the depths of atomic heart,
Where protons and neutrons play their part,
There lies a theory most grand,
Quantum chromodynamics is its name, and

It tells us of the glue that holds,
Quarks and gluons, strong and bold,
Together, in the nucleus confined,
Invisible forces, of energy confined.

With a model, so precise and sure,
It helps us understand the very core,
Of the matter that makes our world,
The secrets, it has, finally unfurled.

From the smallest to the largest scale,
The beauty and the power, it never fails,
To amaze and inspire, with every insight,
Quantum chromodynamics, always in sight.

Hawking radiation

In the depths of space and time,
Where black holes reign supreme,
There's a mystery that still we climb,
With theories of a quantum dream.

The laws of physics don't behave,
Around these objects dark and vast,
So scientists sought to save,
With theories that would stand the test.

And so was born the concept bright,
Of Hawking radiation, born from heat,
Emitting particles of light,
From black holes, both big and small, fleet.

This phenomenon so strange and bold,
Is proof of quantum theory's might,
That particles can leave their hold,
And escape the black hole's fright.

So if you see a speck of light,
In the darkness of the night,
It could be the radiation's sight,
Emanating from a black hole's sight.

And in that moment, you'll take pride,
In the beauty of science's rhyme,

That answers questions we sought to hide,
In the depths of space and time.

Einstein-Podolsky-Rosen paradox.
An entangled dance, of quantum chance,
A paradox woven in its circumstance.
Einstein, Podolsky, Rosen did unite,
In search of the truth, a quantum insight.

Two particles, distant and apart,
Yet their state, together, they impart.
A mystery so grand, so hard to explain,
It challenged the rules of space and brain.

But as they probe, they found a key,
To the door of reality, where it be.
Quantum entanglement, the glue that binds,
A connection so strong, it defies mankind.

So now we know, with this new found art,
That the laws of nature, are so smart.
And as we delve, into this unknown,
The Einstein-Podolsky-Rosen paradox has grown.

Quantum gravity and Theory of everything

The universe is vast, it's mysteries untold,
We search for answers, through stories untold.
With theories of everything, we strive to unite,
The laws of the cosmos, both day and night.

Quantum gravity whispers, a secret untold,
Of how matter and space, and time do hold.
It speaks of the forces, that we can't see,
That shape the fabric, of reality.

With equations and formulas, we strive to find,
A theory that unites, all space and time.
From black holes to stars, and everything in between,
The answers to life, we hope to glean.

As we delve deeper, into the unknown,
We find more questions, yet more to be shown.
With theories of everything, we strive to unite,
The laws of the cosmos, both day and night.

Bell inequality.

In the realm of quantum mechanics,
A paradox did arise,
With entangled particles and such,
A test of truth to prize.

Bell's Inequality came to be,
A measure of correlation,
It showed the world a different view,
Of nature's strange fusion.

Classical thought was challenged then,
And local realism too,
The data showed, beyond a doubt,
Quantum weirdness shines true.

For entangled particles, no matter how far,
Their states are linked in space and time,
And this connection that we cannot see,
Is what gives Bell's Inequality its rhyme.

So now we know that things are not,
What they seem in this great space,
And though it seems like magic,
The science holds its place.

Edward Witten M-theory.

In a world of strings and dimensions vast,
Where theories bend and physics outlasts,
There came a mind that found a new way,
To unify the forces and make them play.

Edward Witten, the man with the plan,
Wrote a theory that took the physics world by stand.
The M-theory, a symphony of beauty and grace,
A new dawn in science, a smile on Einstein's face.

A grand unifying theory, a framework of all,
A map of the universe, from big bang to the fall.
With eleven dimensions, and membranes that swirl,
A dance of the cosmos, a twirl of the whirl.

A bridge between quantum mechanics and gravity,
A missing link, a mystery no more to be.
A song of the universe, a harmony of space,
A symphony of physics, a reflection of the grace.

So let us remember, the man with the mind,
Who wrote a theory, that left us all blind?
For his work shall live on, in this universe so grand,

And guide us to answers, in the search for what is planned.

string theory.

In the quest for answers, in the search for truth,
A theory emerged, a concept uncouth.
The strings of the universe, vibrating with glee,
Each note they play, unlocks the mysteries.

From subatomic particles to the stars so bright,
The strings hold the key, to the mysteries of light.
With eleven dimensions, they weave their tale,
Of how the universe formed, and how it prevails.

Some say it's magic, a tale too wild to be true,
But with each new discovery, the evidence accrues.
For in this theory, the fabric of space and time,
Are woven by strings, in a cosmic rhyme.

A symphony of sounds, in a cosmic dance,
Each movement brings us closer to the chance.
To unlock the secrets, of the universe we see,
With string theory, the possibilities are free.

So listen closely, for the strings they sing,
Of the mysteries of life, and everything.
For in the end, it's just a theory, you see,
But what a theory, it's the theory of everything!

Kepler's laws of planetary motion

The heavens above, in motion so grand,
With planets that orbit, as directed by hand.
Kepler's laws, set forth with such care,
Describe the celestial dance, with precision rare.

The first law states, an ellipse is the path,
The sun at one focus, a concept so math.
The planets they follow, this elliptic way,
A uniform sweep, at equal distance each day.

The second law, a measure of speed,
The closer the planet, the faster it'll proceed.
It's area swept, proportional to time,
A constant ratio, that's truly divine.

The third law, a relationship so grand,
Between the planet's distance and its stand.
The square of the period, is proportional to cube,

Of its average distance, from the sun so huge.

These laws, a testament to Kepler's mind,
With insights so brilliant, of the cosmos combined.
They describe the motion, of planets so bright,
In a dance so graceful, under heaven's light.

Big Bang theory

In the beginning, all was void, a great blackness untold
But then, with a burst, a spark, the universe took hold,
A fireball of immense size, expanding far and wide
And from that explosion, galaxies formed, stars collided.

The fabric of time and space, stretching on and on
An ever-evolving universe, as time goes marching on,
A cosmic dance, a symphony, of particles and light
A story told by the stars, in the endless night.

As matter cooled, and clumps formed, planets began to
take shape
Gravity took hold, and pulled, in its insatiable drape,
And in the vastness of space, Kepler's laws did take root
Guiding planets in their orbits, following the rules.

The Big Bang theory, a tale of creation and strife
A story of a universe, that's come to hold our life,
An endless quest for knowledge, of how it all began
In the hope that we might find, the secrets of the grand plan.

www.ingramcontent.com/pod-product-compliance
Lightning Source LLC
Chambersburg PA
CBHW070421220526
45466CB00004B/1489